PERENNIAL MATH

CONTEST PROBLEMS

2014-2017

Dr. Sylvia Dean was a gifted specialist and administrator for 26 years in Huntsville, Alabama. She created Perennial Math in 2011 upon completion of her doctoral degree. She has hosted many robotics, technology, and mathematics competitions throughout the years and has tirelessly worked on providing educational opportunities to students around the world. It has always been her ambition to leave students with the feeling that "math is cool." These problems are challenging and students should know up front that scoring a 3 out of 6 on any of these tests is extraordinary and above average. Great problem-solvers learn persistence and appreciation for challenge. Good luck as you encourage your students to take the quest to become great mathletes.

ADVANCED
LEVEL

Grades 7-8

Name______________________________

PERENNIAL MATH

TEST # 1

November 2014 Advanced Level

1) If 8 is added to one-third of a number, the result is three times the number. What is the number?

2) Jenny, Bonnie, and Carolyn took turns driving from Milltown to Centerville. First, Jenny drove $\frac{3}{11}$ of the way. Next Bonnie drove 25% of the total distance. Then Carolyn drove the remaining 567 miles. How many miles are in the total distance from Milltown to Centerville?

3) Belinda is twice as old as Rose now. Seven years from now, the sum of their ages will be 44 years. How old will Rose be in 7 years?

4) Mike is sawing logs for firewood. If it takes 12 minutes to saw a log into 3 pieces, at that rate, how many minutes does it take to saw a log into 6 pieces?

5) Two cousins are traveling to their grandmother's house for Thanksgiving. They decide to meet up to ride together. If they live 270 miles apart and one car is traveling at 80 miles per hour and the other car is traveling at 70 miles per hour. How many total minutes before the two cousins meet up?

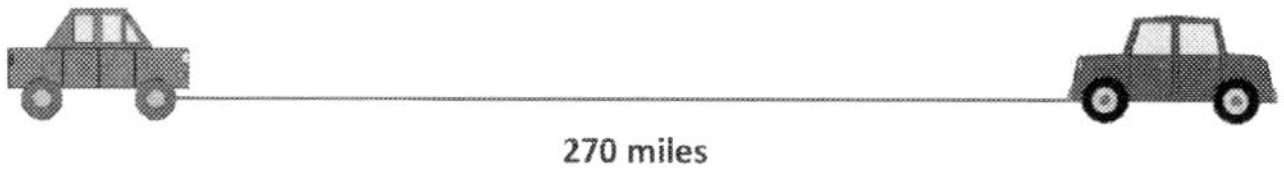

6) 64 votes were cast for Homecoming queen. Barbara received $\frac{1}{4}$ of the votes. Hannah got as many votes as Julie and Donna put together. Donna got $\frac{1}{3}$ as many votes as Julie. How many votes did Julie get?

PERENNIAL MATH

TEST #1

November 2014 Advanced Level Answers

1) If 8 is added to one-third of a number, the result is three times the number. What is the number?	3
2) Jenny, Bonnie, and Carolyn took turns driving from Milltown to Centerville. First, Jenny drove $\frac{3}{11}$ of the way. Next Bonnie drove 25% of the total distance. Then Carolyn drove the remaining 567 miles. How many miles are in the total distance from Milltown to Centerville?	1,188
3) Belinda is twice as old as Rose now. Seven years from now, the sum of their ages will be 44 years. How old will Rose be in 7 years?	17
4) Mike is sawing logs for firewood. If it takes 12 minutes to saw a log into 3 pieces, at that rate, how many minutes does it take to saw a log into 6 pieces?	30
5) Two cousins are traveling to their grandmother's house for Thanksgiving. They decide to meet up to ride together. If they live 270 miles apart and one car is traveling at 80 miles per hour and the other car is traveling at 70 miles per hour. How many total minutes before the two cousins meet up? 	108
6) 64 votes were cast for Homecoming queen. Barbara received $\frac{1}{4}$ of the votes. Hannah got as many votes as Julie and Donna put together. Donna got $\frac{1}{3}$ as many votes as Julie. How many votes did Julie get?	18

PERENNIAL MATH

TEST # 1

November 2014 Advanced Level Solutions

We are only providing some suggested solutions below. There are always multiple methods to solve problems.

1) Let x stand for the number:

$$\frac{x}{3} + 8 = 3x$$

$$\frac{3}{3} + 8 = 3(3)$$

$$1 + 8 = 9$$

3

2) First convert $\frac{3}{11}$ and 25% to like terms so you can know how far Jenny and Bonnie drove. $\frac{3}{11}$ and 25% or $\frac{1}{4}$ have the least common denominator of 44. So they drove

$$\frac{12+11}{44} = \frac{23}{44}$$

Therefore, 567 is equal to $\frac{21}{44}$

$\frac{567}{x} = \frac{21}{44}$ 21x= (567 x 44) = $\frac{24948}{21}$ = 1188

The total distance from Milltown to Centerville is 1,188 miles.

1,188

3) B = 2R (1st equation)
(B+7)+(J+7) = 44 (2nd equation)
Substitute the known value of B from the first equation into the second
(2R+7)+(R+7) = 44
3R+14 = 44 (Subtract 14 from each side)
3R = 30 (Divide each side by 3)
R = 10 (Rose = 10)
If Rose is 10, then Belinda is 20. In 7 years, Rose will be 17 and Belinda will be 27.
17 + 27 = 44

17

PERENNIAL MATH

TEST # 1

November 2014 Advanced Level Solutions

4) To saw a log into 3 pieces it takes 2 cuts which takes 12 minutes. So it takes 6 minutes to make 1 cut. To cut a log into 6 pieces you must make 5 cuts. 5 cuts x 6 minutes = 30 minutes.

30

5) $70t + 80t = 270$

$150t = 270$

$t = \frac{270}{150} = \frac{27}{15} = 1\frac{4}{5}$ hours

	Rate	Time	Distance
Cousin 1	70 mph	t	70t
Cousin 2	80 mph	t	80t
		total distance =	270 miles

1 hour + $\frac{4}{5}$ hours = (60 min + 48 min)

= 108 total minutes to meet up.

108

6) Solve the following equation:

$B + H + J + D = 64$ (Equation 1)

- Barbara got ¼ of the total votes

$B = ¼ * 64 = 16$ (Equation 2)

- So the other 3 girls got 48 votes

$H + J + D = 48$ (Equation 3)

- Hannah got as many votes as Julie and Donna together

$H = J + D$ (Equation 4)

- Donna has $^1/_3$ as many votes as Julie

$J = 3D$ (Equation 5)

- Solve Equation 3 by substituting Equation 4 and Equation 5

$H + J + D = 48$ (Equation 3)

$(3D + D) + 3D + D = 48$

$8D = 48$, so $D = 6$.

- Julie has 3 times as many votes as Donna, so Julie has 18 votes
- Check your work:

$B + H + J + D = 64 \rightarrow 16 + 24 + 18 + 6 = 64$

18

Name________________________________

PERENNIAL MATH

TEST # 2

December 2014 Advanced Level

1) Three friends were comparing the money in their wallets. Alfred and Bob have $65 together. Bob and Charles have $72 together. Alfred and Charles have $79 together. How much money do Alfred, Bob, and Charles have altogether?	
2) In a class of 30 students, 18 have computers, 21 have tablets, and 4 have neither a computer or tablet. How many students in the class have *both* a computer and a tablet?	
3) The ratio of boys to girls at the school pep rally was 8:9. An extra 15 boys came in and the ratio changed to 7:6. How many girls are at the pep rally if there are less than 150 students?	
4) Find three consecutive numbers such that tripling the last number gives the same result as doubling the sum of the first two numbers. What is the third number in this set?	
5) The volume of this cube is 64 cubic inches. How many square inches are in the surface area of the cube? 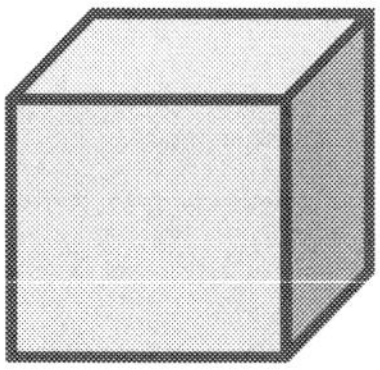	
6) If you listed every possible four-digit number using the digits 2, 3, 4, and 5, what would be the sum of all these four-digit numbers?	

PERENNIAL MATH

TEST #2

December 2014 Advanced Level Answers

1) Three friends were comparing the money in their wallets. Alfred and Bob have $65 together. Bob and Charles have $72 together. Alfred and Charles have $79 together. How much money do Alfred, Bob, and Charles have altogether?	**108**
2) In a class of 30 students, 18 have computers, 21 have tablets, and 4 have neither a computer or tablet. How many students in the class have *both* a computer and a tablet?	**13**
3) The ratio of boys to girls at the school pep rally was 8:9. An extra 15 boys came in and the ratio changed to 7:6. How many girls are at the pep rally if there are less than 150 students?	**54**
4) Find three consecutive numbers such that tripling the last number gives the same result as doubling the sum of the first two numbers. What is the third number in this set?	**6**
5) The volume of this cube is 64 cubic inches. How many square inches are in the surface area of the cube? 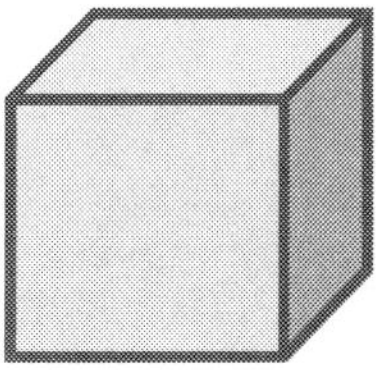	**96**
6) If you listed every possible four-digit number using the digits 2, 3, 4, and 5, what would be the sum of all these four-digit numbers?	**93,324**

PERENNIAL MATH TEST #2

December 2014 Advanced Level Solutions

We are only providing some suggested solutions below. There are always multiple methods to solve problems.

1) Use Algebra.

$$\begin{array}{l} A + B \qquad = 65 \\ \quad B + C = 72 \\ \underline{A + \qquad C = 79} \\ 2A + 2B + 2C = 216 \text{ (divide both sides by 2)} \\ A + B + C = 108 \end{array}$$

108

2) Using a Venn Diagram, we have four Regions in the solution space for this problem. The total of all Regions should equal 30 students. Therefore, we have the following equation:

$$I + II + III + IV = 30$$

Let n = Region II, and substitute in the known values

$$(18 - n) + (n) + (21 - n) + (4) = 30$$

$$43 - n = 30$$

$$13 = n$$

There are 13 students who had both a computer and a tablet .

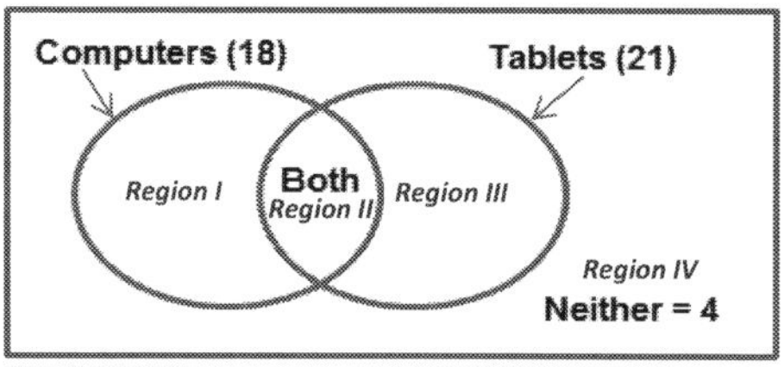

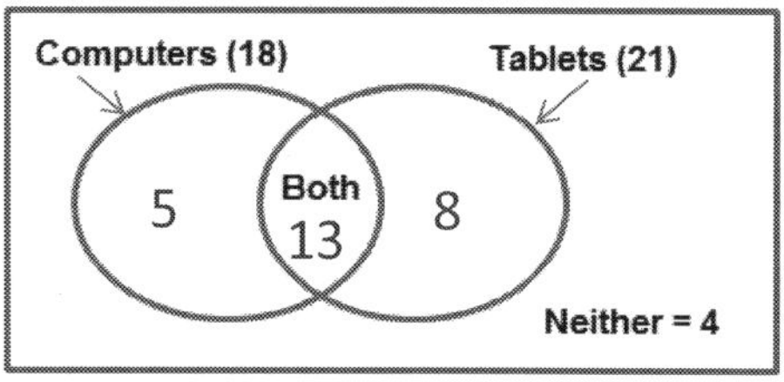

13

3) Originally there were 8*n* boys and 9*n* girls. With an extra 15 boys, we have 8*n*+15 boys and we know the ratio changes to 7:6. Therefore,

$(8n + 15) / 9n = 7/6$ (cross multiply)

$6(8n + 15) = 7 \times 9n$

$48n + 90 = 63n$

$90 = 15n$, hence $n = 6$.

So we can determine that there are $9n = 54$ girls at the pep rally.

54

PERENNIAL MATH

TEST # 2

December 2014 Advanced Level Solutions

4) Find three consecutive numbers such that tripling the last number gives the same result as doubling the sum of the first two numbers. What is the third number in this set?

Use algebra:

The three consecutive numbers can be written as x, x + 1, x + 2

$2 (x + (x + 1)) = 3 (x + 2)$

$2x + 2x + 2 = 3x + 6$

$4x + 2 = 3x + 6$

$x = 4$

The numbers are 4, 5, 6

Check $2(4 + 5) = 3 (6)$ $18 = 18$ ✓

6

5) The volume of a cube is equal to the length of one side cubed. The cube root of 64 is 4 so the cube is 4 inches on each side. The area of one face is length squared or 4x4 = 16 square inches. A cube has 6 faces so the total surface area is 16 x 6 = 96 square inches.

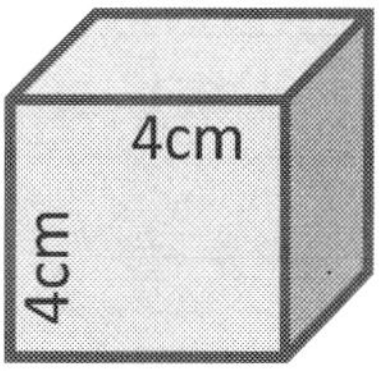

96

6) First make an organized list of all the four-digit numbers starting with the smallest digit and going to the largest. Then get a sum for each list and total the sums.

2345	3245	4235	5234
2354	3254	4253	5243
2435	3425	4325	5324
2453	3452	4352	5342
2534	3524	4523	5423
2543	3542	4532	5432
14664	**20442**	**26220**	**31998**

= 93,324

93,324

Name________________________________

PERENNIAL MATH

TEST #3

January 2015 Advanced Level

1) What is the least natural three-digit number that has three different prime factors?

2) I have a bag containing 47 blue marbles and 53 red marbles. I draw out two marbles. What's the probability that the second marble I draw out is red, given that the first marble I draw is blue? Your answer will be a fraction x/y. What does y equal?

3) Brandon practiced his trombone three consecutive days. Each day he played twice as long as the previous day. If he played 24 minutes on the last day, how many total minutes did he play altogether?

4) One horse eats the same amount as 15 sheep. Two people rent a field together for $185. Person A puts 3 horses in the field. Person B puts 30 sheep in the field. How many dollars should Person A pay based on the amount of grass the animals eat?

5) These three squares are placed side-by-side to form a large rectangle. The perimeter of the large rectangle is 64 centimeters. What is the area of one of the squares?

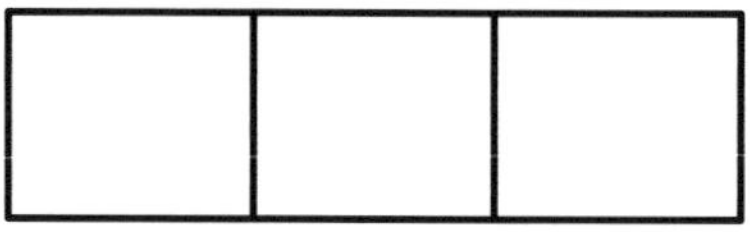

6) Three leprechauns are going to divide 8 pieces of gold. They cannot break any piece of gold. Each one must get at least one gold piece. No two can have the same number of gold pieces. In how many different ways can they divide the gold?

PERENNIAL MATH

TEST #3

January 2015 Advanced Level Answers

1) What is the least natural three-digit number that has three different prime factors?	102
2) I have a bag containing 47 blue marbles and 53 red marbles. I draw out two marbles. What's the probability that the second marble I draw out is red, given that the first marble I draw is blue? Your answer will be a fraction x/y. What does y equal?	99
3) Brandon practiced his trombone three consecutive days. Each day he played twice as long as the previous day. If he played 24 minutes on the last day, how many total minutes did he play altogether?	42
4) One horse eats the same amount as 15 sheep. Two people rent a field together for $185. Person A puts 3 horses in the field. Person B puts 30 sheep in the field. How many dollars should Person A pay based on the amount of grass the animals eat?	111
5) These three squares are placed side-by-side to form a large rectangle. The perimeter of the large rectangle is 64 centimeters. What is the area of one of the squares? 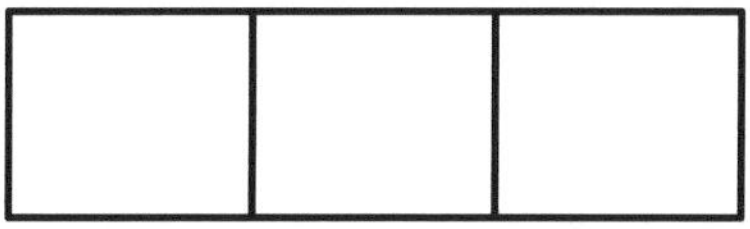	64
6) Three leprechauns are going to divide 8 pieces of gold. They cannot break any piece of gold. Each one must get at least one gold piece. No two can have the same number of gold pieces. In how many different ways can they divide the gold?	12

PERENNIAL MATH TEST #3

January 2015 Advanced Level Solutions

We are only providing some suggested solutions below. There are always multiple methods to solve problems.

1) The least natural three-digit number that has three different prime factors can be found by multiplying the smallest prime factors until you get a three digit number. One other thing you might notice is that 2 x 3 would make the number a multiple of 6.

2 x 3 x 17 = 102

Note: 101 is a prime number so you could start with factoring 102.

102

2) The bag contains 47 blue marbles and 53 red marbles or 100 total marbles. This means the probability of drawing a blue marble out the first time is 47/100. After drawing out the first blue marble then there are 46 blue marbles and 53 red marbles. The probability of drawing a red marble the second time is then 53/99, instead of 53/100, which would be the probability of drawing a red marble the first time. Since x/y = 53/99, y = 99

99

3) Brandon played twice as long as the previous day when he practice for three days. He played 24 minutes on the last day. Set up an equation to see how many total minutes he played altogether. Let X stand for the first day. Then
X + 2X + 4X = ? 4X is equal to 24, so X = 6. Therefore,
6 + 12 + 24 = 42 total minutes of practice.

42

PERENNIAL MATH

TEST #3

January 2015 Advanced Level Solutions

4) Since the ratio of eating is 1 horse to 15 sheep, Peron A has 3 portions compared to Person B's two portions. Since there are 5 total portions, then the rent $185 can be divided by 5 to find out how much one portion costs. One portion costs $37. Since Person A needed 3 portions for the horses, multiply $37 by 3 = $111

111

5) The perimeter of the rectangle is 64cm. There are 8 equal sides going around the shape. So each side of the squares has a length of 64cm/8cm= 8cm. Therefore, the area of one of the small squares is length x width or 8cm x 8cm = 64cm^2.

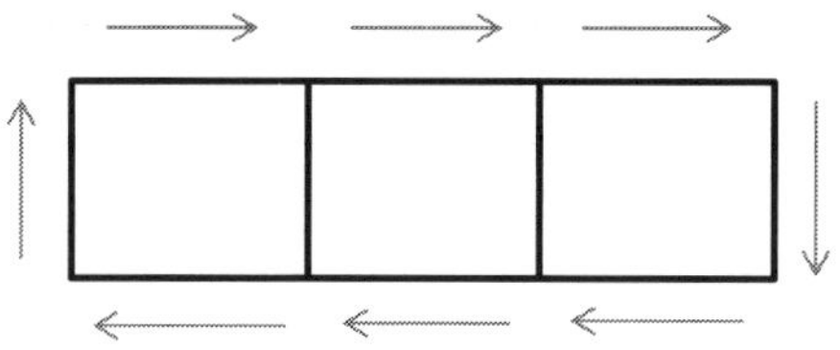

64

6) Make a chart to see the different ways can they divide the gold. There are 12 combinations in which they could divide the gold in order for each leprechaun to receive a different number of gold pieces than the others.

Leprechaun1	Leprechaun2	Leprechaun3
1	2	5
1	3	4
1	4	3
1	5	2
2	1	5
2	5	1
3	1	4
3	4	1
4	1	3
4	3	1
5	1	2
5	2	1

12

Name________________________________

PERENNIAL MATH

TEST #4

February 2015 Advanced Level

1) $(5^2 + (8 \div 4 + 4^2))$

2) Randy has pennies, nickels, and dimes in his bank. How many ways can he combine his coins to make 76 cents?

3) Parents donated fudge for the school fund raiser. Forty pounds of chocolate fudge sold for $2.15 per pound and vanilla fudge sold for $1.90 per pound. Your class made $158.20. How many total pounds of fudge were sold?

4) There are 128 players in a single elimination grand slam tennis tournament. How many games will the champion have to play to win the event?

5) In this problem, the letters H, N, and O stand for different digits. What will the sum NO have to be?

$$\begin{array}{r} OH \\ OH \\ OH \\ +\ OH \\ \hline NO \end{array}$$

6) Michael made a long line of connected squares out of his toothpicks. It took 13 toothpicks to make 4 squares. How many toothpicks would it take to make 99 squares in a line?

PERENNIAL MATH

TEST #4

February 2015 Advanced Level Answers

1) $(5^2 + (8 \div 4 + 4^2))$	**43**
2) Randy has pennies, nickels, and dimes in his bank. How many ways can he combine his coins to make 76 cents?	**72**
3) Parents donated fudge for the school fund raiser. Forty pounds of chocolate fudge sold for $2.15 per pound and vanilla fudge sold for $1.90 per pound. Your class made $158.20. How many total pounds of fudge were sold?	**78**
4) There are 128 players in a single elimination grand slam tennis tournament. How many games will the champion have to play to win the event?	**7**
5) In this problem, the letters H, N, and O stand for different digits. What will the sum NO have to be? OH OH OH + OH NO	**92**
6) Michael made a long line of connected squares out of his toothpicks. It took 13 toothpicks to make 4 squares. How many toothpicks would it take to make 99 squares in a line? 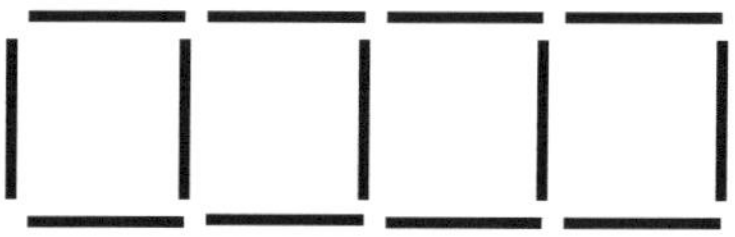	**298**

PERENNIAL MATH

TEST #4

February 2015 Advanced Level Solutions

We are only providing some suggested solutions below. There are always multiple methods to solve problems.

1) Use PEMDAS (Parentheses, Exponents, Multiplication and Division, and Addition and Subtraction) to solve the problem.

$(5^2 + (8 \div 4 + 4^2)) =$

$(5^2 + (8 \div 4 + 16)) =$

$(5^2 + (2 + 16) =$

$(25 + 18) = 43$

43

2) Make a chart to solve the problem.

Dimes	Nickels	Pennies
7	1	1
7		6
6	3	1
6	2	6
6	1	11
6		16
5	5	1
5	4	6
5	3	11
5	2	16
5	1	21
5		26
4	7	1
4	6	6
4	5	11
4	4	16
4	3	21
4	2	26
4	1	31
4		36
3	9	1
3	8	6
3	7	11
3	6	16

Dimes	Nickels	Pennies
3	5	21
3	4	26
3	3	31
3	2	36
3	1	41
3		46
2	11	1
2	10	6
2	9	11
2	8	16
2	7	21
2	6	26
2	5	31
2	4	36
2	3	41
2	2	46
2	1	51
2		56
1	13	1
1	12	6
1	11	11
1	10	16
1	9	21
1	8	26

Dimes	Nickels	Pennies
1	7	31
1	6	36
1	5	41
1	4	46
1	3	51
1	2	56
1	1	61
1		66
	15	1
	14	6
	13	11
	12	16
	11	21
	10	26
	9	31
	8	36
	7	41
	6	46
	5	51
	4	56
	3	61
	2	66
	1	71
		76

72

3) There was 40 lbs. of chocolate fudge sold at $2.15/lb. They made 40 x 2.15 = $86 from the chocolate fudge. If they made a total of $158.20, then subtract $86 and they made $72.20 with vanilla fudge. Since vanilla fudge costs $1.90/lb., they can figure out how much vanilla by dividing 72.20/1.90. They sold 38 lbs. of vanilla fudge. Now add the 40lbs. of chocolate fudge. There was a total of 78 lbs. of fudge sold.

78

PERENNIAL MATH

TEST #4

February 2015 Advanced Level Solutions

4) You can draw out the brackets or you can notice that the number of players is being divided in half after each game. The winner would have to play 7 total games.

Round	Number of Players
1	128
2	64
3	32
4	16
5	8
6	4
7	2

7

5) OH
OH
OH
+ OH
NO

The product of OH x 4 is a 2-digit number. Therefore, O must be either 1 or 2. O has to be even because 4 x H must be even; therefore O is 2. Since O is 2, then 4 x H must be 3 because it is the only number that will yield 3 when multiplied by 4.

OH = 23. Since OH is 23, then 23 x 4 = 92, or NO = 92.

92

6) To do this problem, you could draw out the 99 squares, but the quickest solution would be to notice a pattern.

For every square you are adding 3 more toothpicks. However, if you will notice a relationship in the numbers squares/toothpicks. Let *n* stand for the number of squares. The number of toothpicks for each square can be figured by multiplying the number of squares by 3 and adding 1.
For example: 4 squares would be $3n + 1 = 13$ with $n = 4$
Therefore, to find the number of toothpicks in 99 squares, $99(3) + 1 = 298$ toothpicks.

Number of Squares	Number of toothpicks
4	13
5	16
6	19

298

Name________________________________

PERENNIAL MATH

TEST #1

November 2015 Advanced Level

1) What is the sum of all the prime numbers less than 50 that do not have the digit 3 in them?

2) Bill and Mike were playing marbles. At the end of each game the loser had to give the winner a marble. When they were done, Bill had won 16 times and Mike had 25 more marbles than he started with. How many games did they play?

3) A rectangular garden 60 feet long and 20 feet wide is enclosed by a fence. To make the garden larger, while using the same fence, its shape is changed to a square. By how many square feet does this enlarge the garden?

4) The sum of seven consecutive positive integers is 182. What is the difference between the largest and smallest of these integers?

5) This 3 x 3 grid is made up of a border of eight squares surrounding one square. A 3 x 4 grid would be made up of a border of ten squares surrounding two squares. How many squares would be in the border of a 3 x 100 grid?

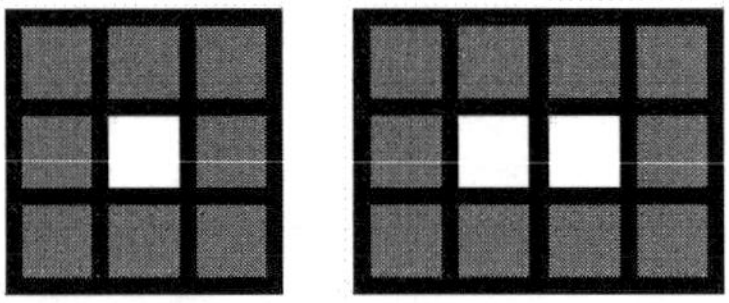

6) A two-digit number *xy* is called "cool" if $x^2 + y^2 = 85$.
How many "cool" numbers are there?

PERENNIAL MATH

TEST # 1

November 2015 Advanced Level Answers

1) What is the sum of all the prime numbers less than 50 that do not have the digit 3 in them?	**178**
2) Bill and Mike were playing marbles. At the end of each game the loser had to give the winner a marble. When they were done, Bill had won 16 times and Mike had 25 more marbles than he started with. How many games did they play?	**57**
3) A rectangular garden 60 feet long and 20 feet wide is enclosed by a fence. To make the garden larger, while using the same fence, its shape is changed to a square. By how many square feet does this enlarge the garden?	**400**
4) The sum of seven consecutive positive integers is 182. What is the difference between the largest and smallest of these integers?	**6**
5) This 3 x 3 grid is made up of a border of eight squares surrounding one square. A 3 x 4 grid would be made up of a border of ten squares surrounding two squares. How many squares would be in the border of a 3 x 100 grid? 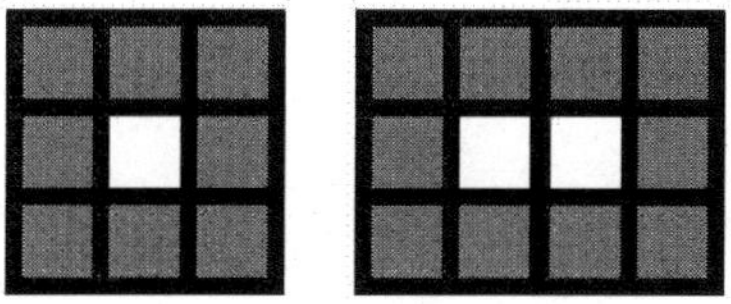	**202**
6) A two-digit number *xy* is called "cool" if $x^2 + y^2 = 85$. How many "cool" numbers are there?	**4**

PERENNIAL MATH

TEST # 1

November 2015 Advanced Level Solutions

We are only providing some suggested solutions below. There are always multiple methods to solve problems.

1) List the prime numbers below 50 and exclude 13, 23, 31, 37, and 43.

$$2 + 5 + 7 + 11 + 17 + 19 + 29 + 41 + 47 = 178$$

178

2) Bill had won 16 times so Mike would have given him 16 marbles. Now Mike would have had to win 16 games to get back to even. Mike also had 25 more marbles than he started with so that is an additional 25 games. So they played 16 + 16 + 25 = 57 games.

57

3) The perimeter of the rectangular garden is 2(60 + 20) = 160 feet. A square with this perimeter has side length of 160/4 = 40 feet. The area of the rectangular garden is 60 x 20 = 1200ft^2 and the area of the square garden is 40 x 40 = 1600ft^2, so the area increases by 400ft^2.

400

PERENNIAL MATH

TEST # 1

November 2015 Advanced Level Solutions

4) The sum of seven consecutive positive integers is 182. Let x stand for the first integer and you can write an equation:

$x + (x+1) + (x+2) + (x+3) + (x+4) + (x+5) + (x+6) = 182$

$7x + 21 = 182$

$7x = 182 - 21$

$7x = 161$

$x = 23$

The integers are 23, 24, 25, 26, 27, 28, and 29

The difference between the largest and smallest of these integers is

$29 - 23 = 6$

6

5) This 3 x 3 grid is made up of a border of eight squares surrounding one square. A 3 x 4 grid would be made up of a border of ten squares surrounding two squares. To find how many squares that would be in the border of a 3 x 100 grid, look for a pattern and use algebra. Let n stand for the number in the sequence that is being multiplied by 3.

Our equation is this: $3 \times n - (n - 2)$

Substitute n = 100: $3 \times (100) - (100 - 2) =$

$300 - 98 = 202$

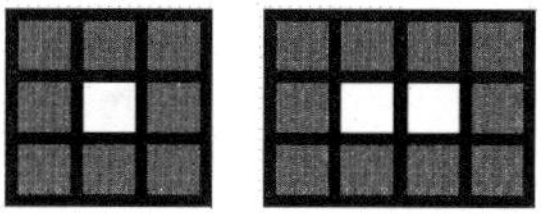

202

6) $x^2 + y^2 = 85$

There are 4 possible numbers for *xy*: 29, 92, 67, and 76.

$2^2 + 9^2 = 4 + 81 = 85$

$6^2 + 7^2 = 36 + 49 = 85$

4

Name________________________________

PERENNIAL MATH

TEST #2

December 2015 Advanced Level

1) Makayla sold half of her doll collection. She then bought seventeen more dolls. She then had 42 dolls. How many dolls did she have to start with?	
2) Over a period of 5 games the basketball team made 44 percent of their shots. If they made between 200 and 240 shots, how many shots did they make?	
3) Andy bought a dozen boxes, each with 24 suckers inside, for $7 each box. He repacked five of these boxes into packages of six suckers each, and sold them for $3 per package. He sold the rest of the suckers separately at the rate of three suckers for $2. How much total profit did he make?	
4) The Panthers beat the Jaguars in the football game. The sum of their scores was 45. The difference of their scores was 27. How much did the Jaguars score?	
5) Place the numbers 1, 2, 3, and 9 in the four shaded regions of the circles so that the sum of the numbers in each circle is the same. Each number must be used only once. What is the sum of the numbers in each circle? 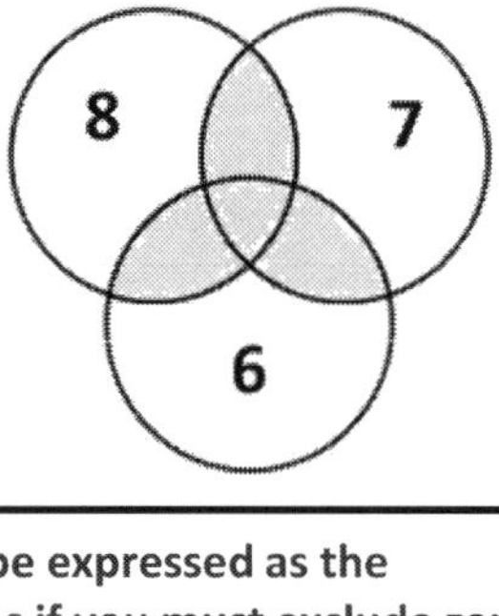	
6) How many whole numbers below 25 can be expressed as the difference of two consecutive perfect squares if you must exclude zero?	

PERENNIAL MATH

TEST #2

December 2015 Advanced Level Answers

<table>
<tr><td>1) Makayla sold half of her doll collection. She then bought seventeen more dolls. She then had 42 dolls. How many dolls did she have to start with?</td><td>50</td></tr>
<tr><td>2) Over a period of 5 games the basketball team made 44 percent of their shots. If they made between 200 and 240 shots, how many shots did they make?</td><td>225</td></tr>
<tr><td>3) Andy bought a dozen boxes, each with 24 suckers inside, for $7 each box. He repacked five of these boxes into packages of six suckers each, and sold them for $3 per package. He sold the rest of the suckers separately at the rate of three suckers for $2. How much total profit did he make?</td><td>88</td></tr>
<tr><td>4) The Panthers beat the Jaguars in the football game. The sum of their scores was 45. The difference of their scores was 27. How much did the Jaguars score?</td><td>9</td></tr>
<tr><td>5) Place the numbers 1, 2, 3, and 9 in the four shaded regions of the circles so that the sum of the numbers in each circle is the same. Each number must be used only once. What is the sum of the numbers in each circle?

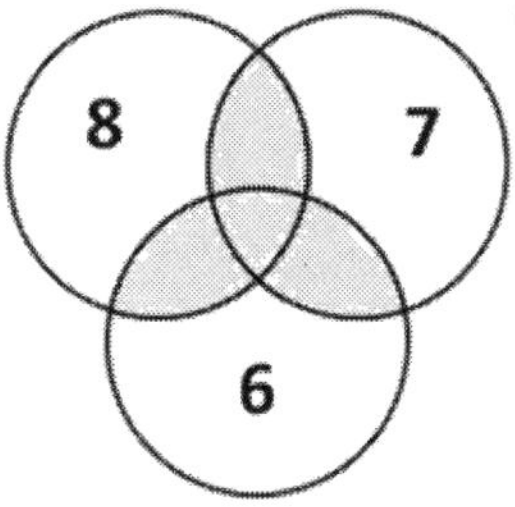

</td><td>20</td></tr>
<tr><td>6) How many whole numbers below 25 can be expressed as the difference of two consecutive perfect squares if you must exclude zero?</td><td>11</td></tr>
</table>

PERENNIAL MATH

TEST # 2

December 2015 Advanced Level Solutions

We are only providing some suggested solutions below. There are always multiple methods to solve problems.

1) Use algebra. Let x stand for the starting number of dolls.

$$\frac{1}{2}x + 17 = 42$$

$$\frac{1}{2}x = 25$$

$$x = 50$$

Makayla started with 50 dolls.

50

2) 44 percent is the same as 44/100. Reduce this and it is 11/25. The only number between 200 and 240 that is divisible by 25 is 225. So they had to have take 225 shots.

225

3) Andy spent 12 x $7 = $84 to buy the suckers.
He separated 5 of the 12 boxes out to sell so that is 5 x 24 =120 suckers.
He repackaged these into 6 per pack so that would have been 120 suckers/6 suckers per pack = 20 packs.
He sold these for $3 per pack, so he made 20 x $3 = $60.
The remaining 7 boxes of 24 suckers was 7 x 24 = 168 suckers.
He packed these in sets of 3 and sold them for $2 each pack.
168 suckers / 3 suckers per pack = 56 packs.
56 packs x $2 = $112.
Now add the amount he made and subtract his original purchase price to find the profit: $60 + $112 = $172.
$172 - $84 = $88 profit.

88

PERENNIAL MATH

TEST # 2

December 2015 Advanced Level Solutions

4) The Panthers beat the Jaguars in the football game. The sum of their scores was 45. The difference of their scores was 27. How much did the Jaguars score?

$$\begin{array}{r} P + J = 45 \\ + \underline{P - J = 27} \\ 2P \quad = 72 \end{array}$$

Therefore, P = 36, so substitute for P: (36) + J = 45

J = 9

9

5) Since the circles have the same sum then the left circle has a sum of 8 + A + D + B and the bottom circle has a sum of 6 + B + D + C.[1] Since B and D are in both circles, and 8 is 2 more than 6, C must be 2 more than A. Of the numbers available to put in the regions, 3 and 1 are the only numbers with a difference of 2.[2]
Now the left circle contains 8 + 1 + B + D, and the right circle contains 1 + D + 3 + 7.[3] Since 1 and D are in both circles, then 8 + B = 7 + 3. Region B = 2. That would leave 9 to go in Region D.[4]

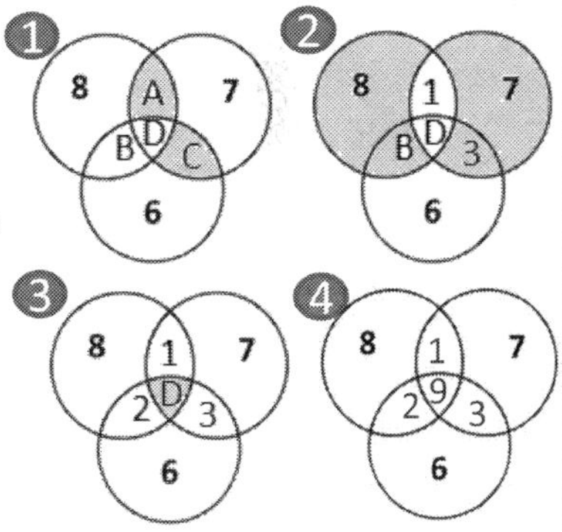

20

6) Make a list of perfect squares: 1, 4, 9, 16, 25, 36, 49, 64, 81, 100, 121, 144. Subtract the consecutive squares to find numbers below 25. A pattern emerges. There are 11 numbers under 25.

144 – 121= 23
121 – 100 = 21
100 – 81 = 19
81 – 64 = 17
64 – 49 = 15
49 – 36 = 13
36 – 25 = 11
25 – 16 = 9
16 – 9 = 7
9 – 4 = 5
4 – 1 = 3

11

Name_______________________________

PERENNIAL MATH

January 2016 Advanced Level

TEST #3

1) What is the difference between the smallest prime number larger than 93 and the largest prime number smaller than 93?	
2) There are 8 boys on the team. If the team can only select 5 boys for the starting line up, how many different starting line ups could be chosen?	
3) Clyde bought a case of 48 helmets for $576. He sold 17 of them for $15 at the flea market, and the rest were sold to a department store for $25 each. How much money was Clyde's profit?	
4) Marvin can mow the lawn in two hours. Greg can mow the same lawn in three hours. If they had two lawn mowers and worked together how long would it take Marvin and Greg to mow the lawn (in minutes)?	
5) Use the numbers 4, 5, 6, 8, 9 to fill in the circles so that each of the three sides along the triangle have a sum of 20. What is the sum of the three corner circles? 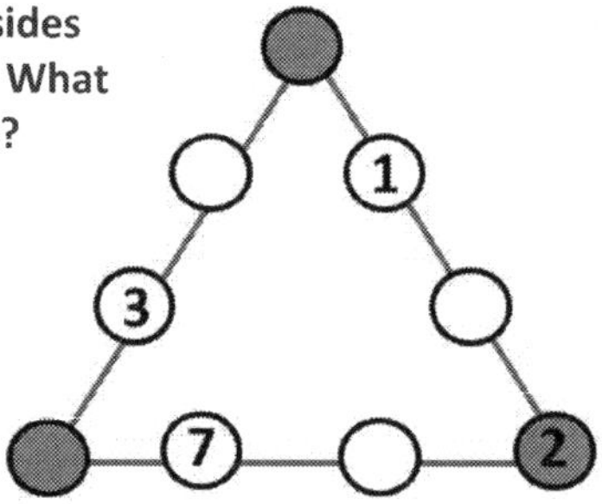	
6) Mike scored 5 points in the first game, 7 points in the second game, 4 points in the third game, and 9 points in the fourth game. How many points does he need to score in the fifth game so that his average number of points per game is 7?	

PERENNIAL MATH

TEST #3

January 2016 Advanced Level Answers

1) What is the difference between the smallest prime number larger than 93 and the largest prime number smaller than 93?	8
2) There are 8 boys on the team. If the team can only select 5 boys for the starting line up, how many different starting line ups could be chosen?	56
3) Clyde bought a case of 48 helmets for $576. He sold 17 of them for $15 at the flea market, and the rest were sold to a department store for $25 each. How much money was Clyde's profit?	454
4) Marvin can mow the lawn in two hours. Greg can mow the same lawn in three hours. If they had two lawn mowers and worked together how long would it take Marvin and Greg to mow the lawn (in minutes)?	72
5) Use the numbers 4, 5, 6, 8, 9 to fill in the circles so that each of the three sides along the triangle have a sum of 20. What is the sum of the three corner circles? 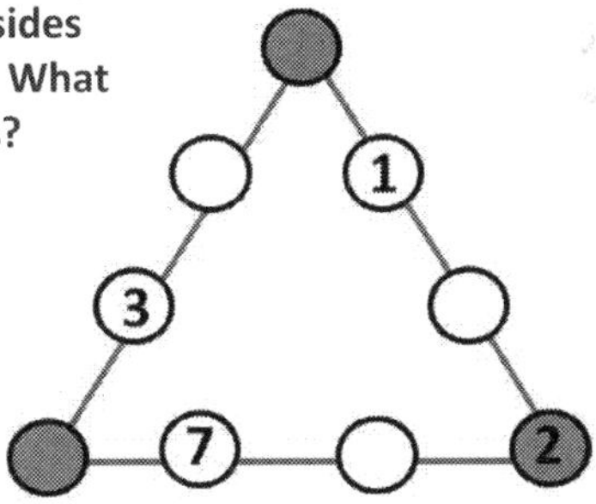	15
6) Mike scored 5 points in the first game, 7 points in the second game, 4 points in the third game, and 9 points in the fourth game. How many points does he need to score in the fifth game so that his average number of points per game is 7?	10

PERENNIAL MATH

TEST # 3

January 2016 Advanced Level Solutions

We are only providing some suggested solutions below. There are always multiple methods to solve problems.

1) The difference between the smallest prime number larger than 93 (97) and the largest prime number smaller than 93 (89) is 8.

8

2) This is a combination problem we are choosing 5 starters from the 8 teammates and the order, position, etc. doesn't matter. The number of combinations of "r" starters chosen from a set of "n" teammates can be expressed as follows:

$$\frac{n!}{(n-r)!r!} = \frac{8!}{(8-5)!\,5!} = \frac{8 \times 7 \times 6 \times 5 \times 4 \times 3 \times 2 \times 1}{(3 \times 2 \times 1)(5 \times 4 \times 3 \times 2 \times 1)} =$$

$\frac{8 \times 7 \times 6}{3 \times 2 \times 1} = \frac{336}{6}$ **= 56 possible team lineups.**

56

3) He spent \$576 for the 48 helmets. He made 17 x \$15 = \$255 selling at the flea market. Next he made \$25 on each of the remaining hats. There were 48 – 17 = 31 hats remaining. 31 x \$25 = \$775. He made a total of \$775 + \$255 = \$1030. Now subtract the original amount he spent to find the profit. \$1030 - \$576 = \$454

454

PERENNIAL MATH

TEST #3

January 2016 Advanced Level Solutions

4) If Marvin can do the entire job in two hours and Greg can do it in three hours, then Marvin can do $^1/_2$ of the job *per hour*, and Greg can do $^1/_3$ *per hour*. To find out how much they can do together *per hour*, add together what they can do individually *per hour*: $^1/_2 + {^1/_3} = {^5/_6}$.
They can do $^5/_6$ of the job *per hour*. Now let "*t*" stand for how long they take to do the job together. Then they can do $^1/_t$ *per hour*, so $^5/_6 = {^1/_t}$. Flip the equation, and you get that $t = {^6/_5}$ = 1.2 hours. Convert this to minutes so that 1.2 hours equals 72 minutes.

72

5) The sum of the corner circles is 8 + 5 + 2 = 15
One way to solve for the missing circles is to make a list of possible sum combinations and see where you can share a number.
For instance:

<u>5</u> + 3 + 4 + 8
<u>5</u> + 7 + 6 + 2

5 is shared and the numbers 6 and 4 are not shared.

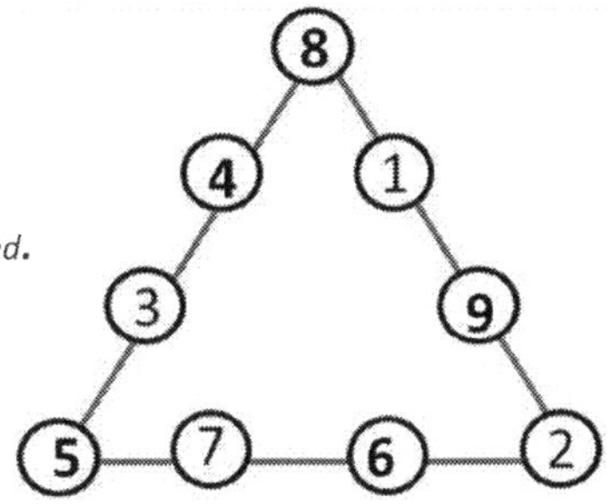

You may also encourage students to use trial and error and make the missing numbers on scraps of paper and move them around on their desktop so that they don't have to erase.

15

6) Average = $\dfrac{\text{Sum of all Items}}{\text{Number of Items}}$

Mike's average score is $7 = \dfrac{5 + 7 + 4 + 9 + X}{5}$

$$35 = 5 + 7 + 4 + 9 + X$$
$$35 - 25 = X$$
$$10 = X$$

He will need to score 10 points.

10

Name______________________________

PERENNIAL MATH

TEST #4

February 2016 Advanced Level

1) An even number between 200 and 300 has 3 different digits. The tens digit is larger that the ones digit. The sum of the three digits is 8 times the hundreds digit. What is the number?

2) Bob has fewer than 6 dozen coins in his piggy bank. If he counts them by twos there is 1 left over. If he counts them by threes, there are 2 left over. Three are left if he counts by fours. Four are left if he counts by fives. How many coins does Bob have?

3) Parents donated 40 dozen cookies for the school fund raiser. Chocolate chip cookies were sold for $2.15 per dozen and sugar cookies were sold for $1.90 per dozen. The total profit made from all cookies was $82.75 How many dozens of chocolate chip cookies were sold?

4) The ratio of Heather's flowers to Belle's flowers was 4 to 3. Heather gave half of her flowers to Belle. She counted her remaining flowers and found that she then had 27 fewer flowers than Belle. How many flowers did Heather have at first?

5) The counting numbers are written across the board in order. There is room for 100 digits across and there are no spaces between each digit. Donna writes two rows of digits. What is the last digit in the second row?

6) The baseball manager was figuring his batting order for his 9 starting players. If his catcher always bats first, how many different batting orders are possible?

PERENNIAL MATH

TEST #4

February 2016 Advanced Level Answers

1) An even number between 200 and 300 has 3 different digits. The tens digit is larger that the ones digit. The sum of the three digits is 8 times the hundreds digit. What is the number?	286
2) Bob has fewer than 6 dozen coins in his piggy bank. If he counts them by twos there is 1 left over. If he counts them by threes, there are 2 left over. Three are left if he counts by fours. Four are left if he counts by fives. How many coins does Bob have?	59
3) Parents donated 40 dozen cookies for the school fund raiser. Chocolate chip cookies were sold for $2.15 per dozen and sugar cookies were sold for $1.90 per dozen. The total profit made from all cookies was $82.75 How many dozens of chocolate chip cookies were sold?	27
4) The ratio of Heather's flowers to Belle's flowers was 4 to 3. Heather gave half of her flowers to Belle. She counted her remaining flowers and found that she then had 27 fewer flowers than Belle. How many flowers did Heather have at first?	36
5) The counting numbers are written across the board in order. There is room for 100 digits across and there are no spaces between each digit. Donna writes two rows of digits. What is the last digit in the second row? 12345678910111213...	0
6) The baseball manager was figuring his batting order for his 9 starting players. If his catcher always bats first, how many different batting orders are possible?	40,320

PERENNIAL MATH TEST #4

February 2016 Advanced Level Solutions

We are only providing some suggested solutions below. There are always multiple methods to solve problems.

1) An even number between 200 and 300 has 3 different digits. The tens digit is larger that the ones digit. The sum of the three digits is 8 times the hundreds digit. What is the number?

The hundreds digit must be 2 since the number is between 200-300.
The sum of the digits is 8 times the hundreds digit so the sum must be 8 x 2 = 16.
Since the tens digit is larger than the ones digit and the number is even, the tens digit must be 8 and the ones digit 6 so that the three numbers adds up to 16.
2 + 8 + 6 = 16. The three digit number is 286.

286

2) Make a chart to compare the piles. The number must be odd since there is one left over when Bob puts the coins in piles of 2. There are 59 coins in the piggy bank because it is less than 72 coins and fulfills all the other requirements as shown in the chart.

piles of 3 with 2 left over	5	8	11	14	17	20	23	26	29	32	35	38	41	44	47	50	53	56	59
piles of 4 with 3 left over	7	11	15	19	23	27	31	35	39	43	47	51	55	59	63	67	71		
piles of 5 with 4 left over	9	14	19	24	29	34	39	44	49	54	59	64	69						

59

3) Make a table starting with 20 of each type of cookie. The total profit must increase so increase the number of chocolate chip cookies. Notice that every increase of chocolate chip adds a quarter profit. To get from $81 profit to $82.75, you would need to increase $1.75 which is 7 quarters. So increase the number of chocolate chip dozens to 27 and add up your totals. There were 27 dozens of chocolate chip cookies sold.

Dozens of Chocolate Chip Cookies	20	21	22	23	24	25	26	27
Dozens of Sugar Cookies	20	19	18	17	16	15	14	13
Ch. Chip Dozens x $2.15	43	45.15	47.3	49.45	51.6	53.75	55.9	58.05
Sugar Dozens x $1.90	38	36.1	34.2	32.3	30.4	28.5	26.6	24.7
Total Amount (must = $82.75)	81	81.25	81.5	81.75	82	82.25	82.5	82.75

27

PERENNIAL MATH

TEST #4

February 2016 Advanced Level Solutions

4) The ratio of Heather's flowers to Belle's flowers was 4 to 3.

Heather gave half of her flowers to Belle. She counted her remaining flowers and found that she then had 27 fewer flowers than Belle. Each portion is worth 9. Since Heather had 4 portions to start with she 9 x 4 = 36 flowers to start with.

36

5) You can list out the numbers across your page

1234567891011121314151617181920212223242526272829303132333435363738394041424344454647484950515253545

5565758596061626364656667686970717273747576777879808182838485868788899091929394959697989910010110210

Or you can count the number of digits by grouping.

<u>1st Row</u>

Single digit #'s 1 through 9 = 9 digits

Double digit #'s 10 through 55 = 92 digits

= 101 digits, so we have to split the last number, "55", between the two rows

<u>2nd Row</u>

Single digit #'s 5 (one's digit of 55) = 1 digit

Double digit #'s 56 through 99 = 88 digits

= 89 digits through the number 99

We need 100 – 89 = 11 more digits, so start counting at 100 until you get 11 digits.

100 101 102 10

"0" is the last digit on the 2nd Row

0

6) The first position stays the same so use the permutation 8 x 7 x 6 x 5 x 4 x 3 x 2 x 1. Since there are 8 possible players for the second position, 7 for the third (1 is taken off), 6 for the fourth (1 is taken off), etc.

So there are 40,320 different possible batting orders.

40,320

Name________________________________

PERENNIAL MATH

November 2016 Advanced Level

TEST #1

1) a + b + c + d = 33. a + c = 21, and a + b + d = 21. What is the value of a?	
2) How much less interest is earned at 6% simple interest on a $10,000 principal over 4 years than a 9% interest for 8 years?	
3) Michael has lived in Denver for *n* years. If he stays there another 30 years he will have lived there n^2 years. How many years has Michael lived there?	
4) A farmer intends to use 280 meters of fence to construct rectangular pens with length L and width W. He plans to build the boundary fences plus one internal dividing fence (see the diagram). If he lets the width be 40 meters, the total area enclosed will be A square meters. If he lets the width be 48 meters, the total area enclosed will be B square meters. What is the difference in B – A in square meters? 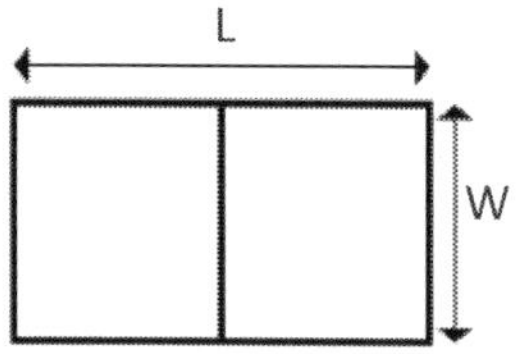	
5) Between 1:00 a.m. and 8:00 a.m. how many times will the digital clock display a three-digit number that is a perfect square?	
6) A survey on a sample of 25 new cars being sold at a local dealership found that 15 had navigation system; 2 had navigation system and power windows but no Bluetooth; 12 had bluetooth; 6 had navigation system and bluetooth but no power windows; 11 had power windows; 4 had bluetooth and power windows; and3 had all three options. What is the number of cars that had none of the three options?	

PERENNIAL MATH

TEST #1

November 2016 Advanced Level Answers

1) a + b + c + d = 33. a + c = 21, and a + b + d = 21. What is the value of a?	9
2) How much less interest is earned at 6% simple interest on a $10,000 principal over 4 years than a 9% interest for 8 years?	4800
3) Michael has lived in Denver for *n* years. If he stays there another 30 years he will have lived there n^2 years. How many years has Michael lived there?	6
4) A farmer intends to use 280 meters of fence to construct rectangular pens with length L and width W. He plans to build the boundary fences plus one internal dividing fence (see the diagram). If he lets the width be 40 meters, the total area enclosed will be A square meters. If he lets the width be 48 meters, the total area enclosed will be B square meters. What is the difference in B – A in square meters? 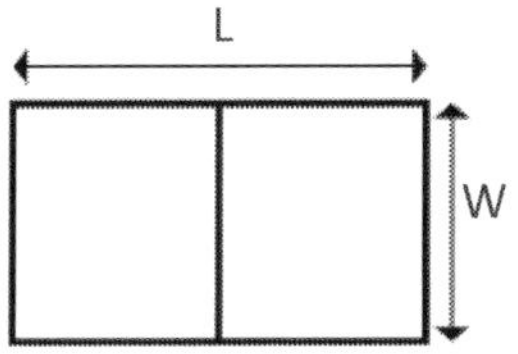	64
5) Between 1:00 a.m. and 8:00 a.m. how many times will the digital clock display a three-digit number that is a perfect square?	10
6) A survey on a sample of 25 new cars being sold at a local dealership found that 15 had navigation system; 2 had navigation system and power windows but no bluetooth; 12 had bluetooth; 6 had navigation system and bluetooth but no power windows; 11 had power windows; 4 had bluetooth and power windows; and3 had all three options. What is the number of cars that had none of the three options?	2

PERENNIAL MATH

TEST # 1

November 2016 Advanced Level Solutions

We are only providing some suggested solutions below. There are always multiple methods to solve problems.

1) a + b + c + d = 33. a + c = 21, and a + b + d = 21. To find the value of a subtract equations and solve for the unknowns. What is the value of a?

$$\begin{array}{r} a + b + c + d = 33 \\ - \underline{\;a + \quad\; c \quad\; = 21} \\ b + \quad d = 12 \end{array}$$

Now substitute this in to a + b + d = 21. a + 12 = 21, Therefore a = 9.

9

2) The interest earned at 6% simple interest on a $10,000 principal over 4 years is (10,000 x .06) x 4 = $2400. The interest earned at 9% interest for 8 years is (10,000 x .09) x 8 = $7200. $7200 - $2400 - $4800.

4800

3) Create a table to see the relationship of n. Michael has lived in Denver for 6 years because 6 + 30 = 36 which is 6^2.

n	n^2	n + 30
1	1	21
2	4	22
3	9	23
4	16	24
5	25	25
6	36	36

6

PERENNIAL MATH

TEST #1

November 2016 Advanced Level Solutions

4) To determine the area for both scenarios, solve for length of the pen and use the L * W = Area equation. Note that there are 3 widths for this fence layout. So, the amount of fence would be 3(L) + 2(W) = 280. For scenario A the length would be 80 meters since 3(40) + 2(80) = 280 meters and the area would be 40 x 80 = $3200m^2$. If he lets the width be 48 meters in scenario B, then the length would be 68 meters since 3(48) + 2 (68) = 280 meters and the area would be 48 x 68 = $3456m^2$. The difference of B - A = $3264m^2 - 3200m^2 = 64m^2$.

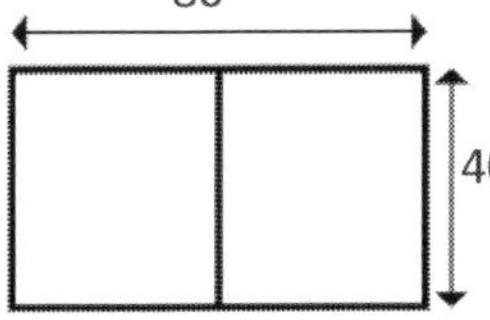

Area A = $3200m^2$

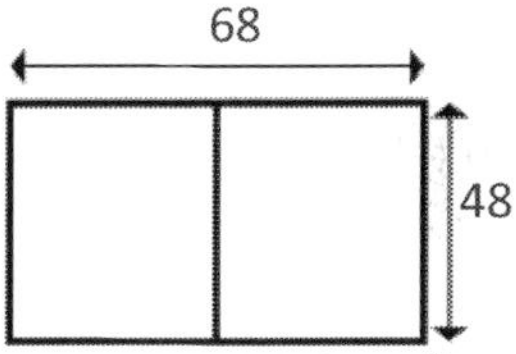

Area B = $3264m^2$

64

5) Between 1:00 a.m. and 8:00 a.m. the digital clock displays numbers that are perfect squares. The numbers are 1:21, 1:44, 2:25, 2:56, 3:24, 4:00, 4:41, 5:29, 6:25, 7:29. Therefore, the digital clock display a three-digit number that is a perfect square 10 times?

10

6) Use a Venn Diagram to solve the problem. Total Number of cars according to the diagram = 2 + 6 + 3 + 1 + 5 + 2 + 4 = 23. Therefore, number of cars having none of the given options = 25 - 23 = 2.

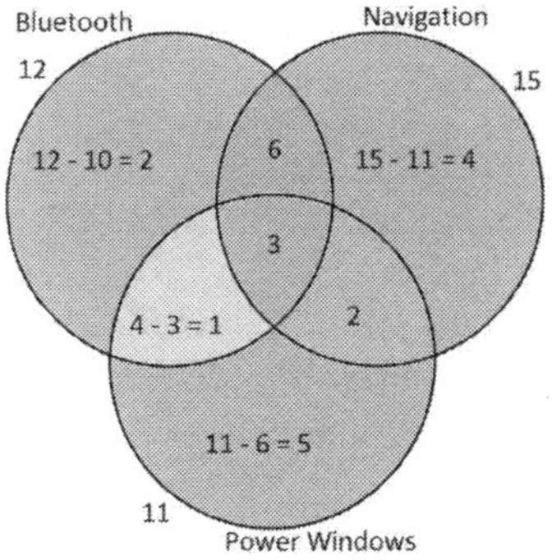

2

Name________________________________

PERENNIAL MATH

TEST #2

December 2016 Advanced Level

1) $9^2 - 8^2 + 7^2 - 6^2 + 5^2 =$	
2) Function Machine A takes a number and squares it and then adds one to the result. (For instance, if 3 is put into Machine A, 10 will be the output.) Function Machine B takes a number divides it by 2 and then adds 5 to the result. (For instance, if 3 is put into Machine B, 6.5 will be the output.) So, a number is put into Machine A; the output from A is put into Machine B; 30 is the output from Machine B. What was the original number that was put into Machine A?	
3) The average of two numbers A and B is 7. The average of three numbers C, D, and E is 12. What is the average of all five numbers?	
4) If $b \bigtriangleup c = b \times c + c^2$ then what is the value of $(6 \bigtriangleup 4) \bigtriangleup 2$	
5) Sam went into a store and spent half of his money and then $15 more. He went into a second store and spent half of his remaining money and then $25 more. Then he had no money left. How much money did Sam have when he went into the first store?	
6) Find the sum of the largest prime factor of 462, the largest prime factor of 195, and the largest prime factor of 306.	

PERENNIAL MATH

TEST #2

December 2016 Advanced Level Answers

1) $9^2 - 8^2 + 7^2 - 6^2 + 5^2 =$	55
2) Function Machine A takes a number and squares it and then adds one to the result. (For instance, if 3 is put into Machine A, 10 will be the output.) Function Machine B takes a number divides it by 2 and then adds 5 to the result. (For instance, if 3 is put into Machine B, 6.5 will be the output.) So, a number is put into Machine A; the output from A is put into Machine B; 30 is the output from Machine B. What was the original number that was put into Machine A?	7 or -7
3) The average of two numbers A and B is 7. The average of three numbers C, D, and E is 12. What is the average of all five numbers?	10
4) If $b \bigtriangleup c = b \times c + c^2$ then what is the value of $(6 \bigtriangleup 4) \bigtriangleup 2$	84
5) Sam went into a store and spent half of his money and then $15 more. He went into a second store and spent half of his remaining money and then $25 more. Then he had no money left. How much money did Sam have when he went into the first store?	130
6) Find the sum of the largest prime factor of 462, the largest prime factor of 195, and the largest prime factor of 306.	41

PERENNIAL MATH

TEST # 2

December 2016 Advanced Level Solutions

We are only providing some suggested solutions below. There are always multiple methods to solve problems.

1) $9^2 - 8^2 + 7^2 - 6^2 + 5^2 = 81 - 64 + 49 - 36 + 25 = 55$

55

2) Work backwards. Since the final number from Machine B is 30, solve for the Machine A output (A_{output}) using the following Machine B processing:

$(A_{output} \div 2) + 5 = B_{output}$

$(A_{output} \div 2) + 5 = 30$ (subtract 5 from both sides)

$(A_{output} \div 2) = 25$ (multiply by 2 on both sides)

$A_{output} = 50$

Now solve for the original number that was put into Machine A (A_{input}).

$(A_{input})^2 + 1 = A_{output}$

$(A_{input})^2 + 1 = 50$ (subtract 1 from both sides)

$(A_{input})^2 = 49$ (take the square root of both sides)

$A_{input} = 7$

7 or -7

3) The average of A and B is 7 so their total sum is 14. The average of C, D, and E is 12 so their sum is 36. So, the total sum for both averages is 14 + 36 = 50. Now divide the sum by 5 (the number of items), and 50/5 = 10. The average of the 5 numbers is 10.

10

PERENNIAL MATH

TEST #2

December 2016 Advanced Level Solutions

4) If b ⌂ $c = b \times c + c^2$. Then (6 ⌂ 4) ⌂ 2 can be solved by first solving the parenthesis applying the formula supplied. Then complete the problem by applying the formula to the remaining problem using the answer to the parenthesis and 2.

(6 ⌂ 4) ⌂ 2 =
($b \times c + c^2$) ⌂ 2 =
($6 \times 4 + 4^2$) ⌂ 2 =
(24 + 16) ⌂ 2 =

40 ⌂ 2 =
$40 \times 2 + 2^2 =$
$80 + 4 = 84$

84

5) Draw this out: Sam went into the first store and spent a half of his money and then \$15 more.

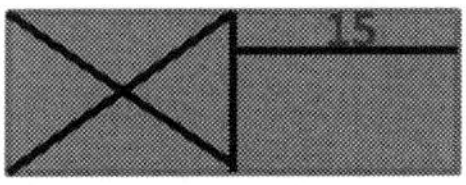

He went into a second store and spent half of his remaining money and then \$25 more. Then he had no money left.

that portion = \$25

So ½ of the total money = \$15 + \$25 + \$25 = \$65, so he had \$65 x 2 = \$130 when he went into the first store.

130

6) To find the sum of the largest prime factor of 462, the largest prime factor of 195, and the largest prime factor of 306, make a factor tree of each number and add up the largest prime factor of each.

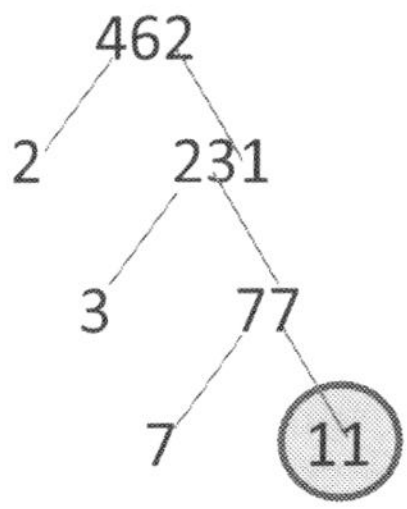

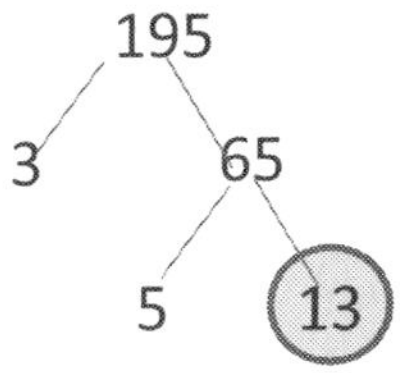

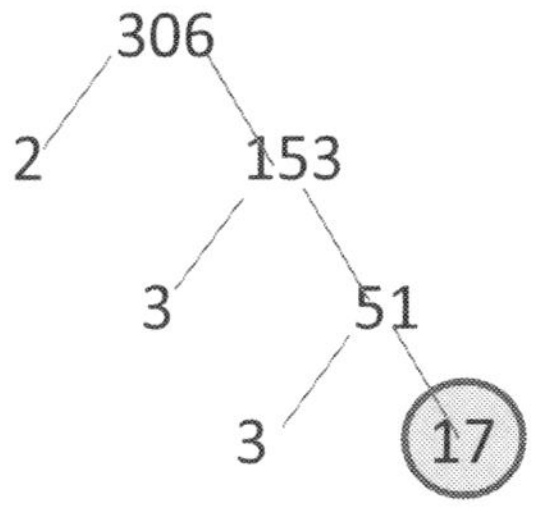

The sum is 11 + 13 + 17 = 41

41

Name________________________________

PERENNIAL MATH

TEST #3

January 2017 Advanced Level

1) Val's weekly allowance increased by 50% up to $27. What was the original amount of her weekly allowance?	
2) Ben and Lucy have a combined weight of 221 pounds. Ben and Tyler have a combined weight of 229 pounds. Lucy and Tyler have a combined weight of 218 pounds. How many pounds does Lucy weigh?	
3) A 4 oz. bottle of sunscreen sells for $4.80. The 8 oz. bottle sells for 60% less per ounce than the 4 oz. bottle. How much does five 8 oz. bottles of sunscreen cost?	
4) Given: $a \# b = \frac{a^2+b^2}{ab}$. Evaluate: (2 # 4) + (4 # 8)	
5) The ratio of video game points for Jeff and Bob was 7 to 2. The total number of points scored was 1971. How many more points did Jeff score than Bob?	
6) Mr. Jones paid a total of $854 (including tax) for two airline tickets, one for an adult and the other for a child. If the airline gives a 25% discount on the child's ticket and the airport charges a $21 tax per ticket regardless of type, what is the total cost of the child's ticket (including the airport tax)?	

PERENNIAL MATH

TEST #3

January 2017 Advanced Level Answers

1) Val's weekly allowance increased by 50% up to \$27. What was the original amount of her weekly allowance?	18
2) Ben and Lucy have a combined weight of 221 pounds. Ben and Tyler have a combined weight of 229 pounds. Lucy and Tyler have a combined weight of 218 pounds. How many pounds does Lucy weigh?	105
3) A 4 oz. bottle of sunscreen sells for \$4.80. The 8 oz. bottle sells for 60% less per ounce than the 4 oz. bottle. How much does five 8 oz. bottles of sunscreen cost?	19.20
4) Given: $a \# b = \frac{a^2+b^2}{ab}$. Evaluate: (2 # 4) + (4 # 8)	5
5) The ratio of video game points for Jeff and Bob was 7 to 2. The total number of points scored was 1971. How many more points did Jeff score than Bob?	1,095
6) Mr. Jones paid a total of \$854 (including tax) for two airline tickets, one for an adult and the other for a child. If the airline gives a 25% discount on the child's ticket and the airport charges a \$21 tax per ticket regardless of type, what is the total cost of the child's ticket (including the airport tax)?	369

PERENNIAL MATH

TEST # 3

January 2017 Advanced Level Solutions

We are only providing some suggested solutions below. There are always multiple methods to solve problems.

1) Val's weekly allowance increased by 50% up to $27. Use algebra and let the term x be her original allowance.

$$x + .5x = 27$$
$$1.5x = 27$$
$$x = 27/1.5$$
$$x = 18$$

18

---

2) Ben and Lucy have a combined weight of 221 pounds. Ben and Tyler have a combined weight of 229 pounds. Lucy and Tyler have a combined weight of 218 pounds. How many pounds does Lucy weigh? Use algebra.

$$\begin{aligned} B + L \quad\quad &= 221 \\ B + \quad\quad T &= 229 \\ \underline{\quad L + T} &= \underline{218} \\ 2B + 2L + 2T &= 668 \end{aligned}$$ (divide both sides by 2)

$$B + L + T = 334$$

Subtract out the original equation B + T to solve for L.

$$(B + T + L) - (B + T) = L$$
$$334 - 229 = 105$$

105

---

3) A 4 oz. bottle of sunscreen sells for $4.80.
So each ounce sells for 4.80/4 = $1.20.
60% less than $1.20 is $1.20 *.4 = .48 per ounce.
Therefore one 8 oz. bottle cost 8 x .48 = $3.84.
Five 8 oz. bottles would cost 3.84 x 5 = $19.20

19.20

PERENNIAL MATH

TEST #3

January 2017 Advanced Level Solutions

4) Given: $a \# b = \frac{a^2+b^2}{ab}$. Evaluate: (2 # 4) + (4 # 8)

Given: $2 \# 4 = \frac{2^2+4^2}{2*4} = \frac{4+16}{8} = \frac{20}{8} = 2.5$

Given: $4 \# 8 = \frac{4^2+8^2}{4*8} = \frac{16+64}{32} = \frac{80}{32} = 2.5$

Therefore: (2 # 4) + (4 # 8) = 2.5 + 2.5 = 5

5

5) The ratio of video game points for Jeff and Bob was 7 to 2. The total number of points scored was 1971. How many more points did Jeff score than Bob? There are 7 + 2 = 9 total portions. Find out how much one portion is by dividing 1971/9 = 219. Therefore Jeff scored 7 x 219 = 1533 points. Bob scored 2 x 219 = 438.
So, Jeff scored 1533 – 438 = 1095 more points than Bob.

1,095

6) Mr. Jones paid a total of $854 (including tax) for two airline tickets, one for an adult and the other for a child. If the airline gives a 25% discount on the child's ticket and the airport charges a $21 tax fee per ticket regardless of type, what is the total cost of the child's ticket (including the airport tax)?
Work backwards $854 - $42 tax = $812 for the two tickets.
If x represents the price of the adult ticket, then the child ticket is 3/4x

x + 3/4x = 812

7/4x = 812

x = 812 ▪ 4/3 = $464 (adult ticket price)

Since the adult ticket (x) costs 25% more, the child ticket costs

x * .75 =

$464 * .75 = $348.

So the total cost of the child ticket is $348 + $21 tax = $369

369

Name________________________________

PERENNIAL MATH

TEST #4

February 2017 Advanced Level

1) When three consecutive whole numbers are added, the total is 54. What is the product of the highest and lowest number?

2) How many numbers from 1-100 have exactly 3 factors?

3) At a pep rally, one-half of the students wore blue, one-third of the students wore gold, fifteen people wore neither color, and nobody wore both colors. How many people were at the pep rally?

4) If the sequence here were to be continued with each square continuing to grow and keep the same pattern, how many small black squares would be in the 20th term?

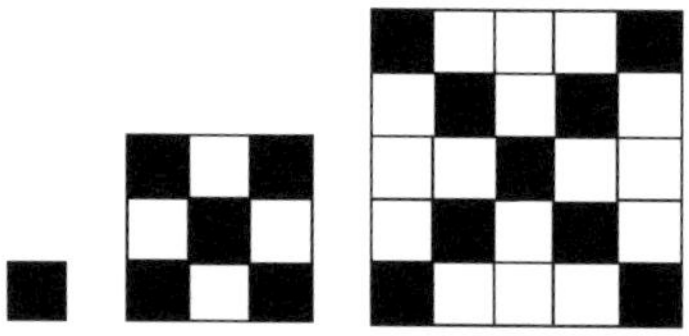

5) Billy has some play money with \$1, \$2 and \$3 bills. There are 11 more \$2 bills than \$1 bills. There are 18 fewer \$3 bills than \$1 bills. The total value of all bills is \$100. What is the value of the \$1 and \$3 bills together?

6) What is the sum of the numbers inserted in the two boxes to make these equations true?

$$48 = 10 - \square = \frac{\square}{1/2}$$

PERENNIAL MATH TEST #4

February 2017 Advanced Level Answers

1) When three consecutive whole numbers are added, the total is 54. What is the product of the highest and lowest number?	**323**
2) How many numbers from 1-100 have exactly 3 factors?	**4**
3) At a pep rally, one-half of the students wore blue, one-third of the students wore gold, fifteen people wore neither color, and nobody wore both colors. How many people were at the pep rally?	**90**
4) If the sequence here were to be continued with each square continuing to grow and keep the same pattern, how many small black squares would be in the 20th term? 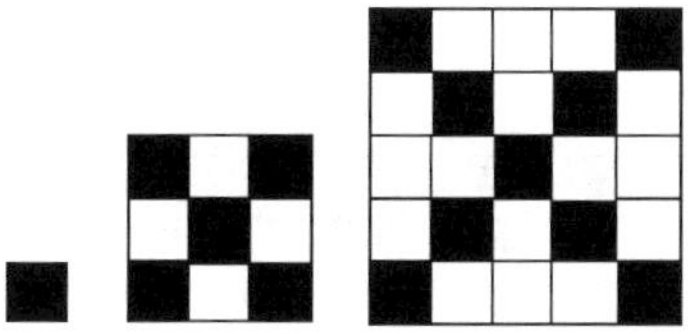	**77**
5) Billy has some play money with \$1, \$2 and \$3 bills. There are 11 more \$2 bills than \$1 bills. There are 18 fewer \$3 bills than \$1 bills. The total value of all bills is \$100. What is the value of the \$1 and \$3 bills together?	**34**
6) What is the sum of the numbers inserted in the two boxes to make these equations true? $48 = 10 - \square = \frac{\square}{1/2}$	**-14**

PERENNIAL MATH

TEST #4

February 2017 Advanced Level Solutions

We are only providing some suggested solutions below. There are always multiple methods to solve problems.

1) First find the three consecutive whole numbers are added, the total is 54 by dividing the total by 3 to find the middle number. The middle number is 18, so the three numbers are 17, 18, and 19. The product of the highest and lowest number is 17 x 19 = 323.

323

2) How many numbers from 1-100 have exactly 3 factors? Only numbers that are squares have three factors, but not all square numbers. The number 4 has factors: 1, 2 and 4. The other square numbers that have three factors are 9, 25, and 49. 4 is the only even number with three factors. Therefore there are only four numbers between 1-100 that have three factors.

4

3) At a pep rally, one-half of the students wore blue, one-third of the students wore gold, fifteen people wore neither color, and nobody wore both colors. To solve for the number of students at the pep rally we know that 1/2 + 1/3 + 15 = the total number of students. So convert the fractions to like terms. So 3/6 + 2/6 = 5/6 (wore blue & gold). Therefore, we know that the 15 students wearing neither color has to be 1/6 of the students. So 6 x 15 = 90 students at the pep rally.

90

PERENNIAL MATH

TEST # 4

February 2017 Advanced Level Solutions

4) If you make a table you will notice that you add 4 black squares to each new term in the sequence. So you could start with the 9 squares in the 3rd term and just add (17 x 4) to 9 to give you the squares you would have in the 20th term of the sequence. There are 77 squares in the 20th term.

Position	1	2	3	4	5	6	7	8	9	10	11	12	13	14	15	16	17	18	19	20
Black Squares	1	5	9	13	17	21	25	29	33	37	41	45	49	53	57	61	65	69	73	77

77

5) Use algebra and let n represent the number of $1 bills.

1n + 2(n+11) + 3(n-18) = 100

n + 2n + 22 + 3n - 54 = 100

6n = 132

n = 22 (number of $1 bills)

then number of $2 bills:

n + 11 = 22 + 11 = 33

then number of $3 bills

n - 18 = 22 – 18 = 4

So 22 $1 bills, 33 $2 bills and 4 $3 bills

Check the total value: $22 + $66 + $12 = $100

The value of just the $1 plus the $3 bills is $22 + $12 = $34

34

6) What is the sum of the numbers inserted in the two boxes to make these equations true?

$$48 = 10 - \square = \frac{\square}{1/2}$$

48 = 10 – (-***38***) for the 1st box

48 = ***24*** ÷ 1/2 for the 2nd box

Therefore the sum of the two numbers in the boxes is -38 + 24 = -14

-14

Made in the USA
Monee, IL
05 November 2026

67811035R00031